A Cat's Body

BY JOANNA COLE

Photographs by Jerome Wexler

William Morrow and Company
New York   1982

Library of Congress Cataloging in Publication Data

Cole, Joanna.
    A cat's body.
    Summary: Photographs and text show how the anatomy
of the common house cat, from limber backbone to retractable claws,
equips it to be a hunter of small rodents.
    1. Cats—Juvenile literature. 2. Cats—Behavior—Juvenile literature.
3. Cats—Anatomy—Juvenile literature. [1. Cats. 2. Predatory animals]
I. Wexler, Jerome, ill.   II. Title.   SF445.7.C64   599.74'428   81-22386
ISBN 0-688-01052-0   ISBN 0-688-01054-7 (lib. bdg.)   AACR2

For her helpful reading of the manuscript,
the author thanks Debra Pirotin, D.V.M.,
a veterinarian who treats cats
in Manhattan and Westchester County, New York.

For Connie Epstein,
with love and gratitude.

Cats are graceful and quick, playful and curious. These special qualities make them beautiful and fun to have as pets. But these same traits originally developed in cats to make them better hunters.

Today's house cats are descended from African wild cats that had to hunt to stay alive. A pet cat's dinner may come from the supermarket, but much of its nature is inherited from its wild ancestors. And its body is still a hunter's body, designed by nature to capture and kill small rodents.

The cat's special skeleton makes its body extra limber and flexible. Because of its small collarbones and narrow chest, the cat can slip through tight places. Its forefeet almost step in each other's tracks, so it can walk easily on a path only two inches wide. Creeping through underbrush, it can come up silently on unsuspecting prey.

A cat also has a super flexible spine. Unlike most animals, it can arch its back into a half circle. This flexible backbone lets a cat turn and twist in pursuit of prey. It also gives the animal a longer stride, since the back flexes as the cat runs. So, for a short distance, a cat can go faster than a dog of the same size.

The cat's powerful muscles work with its flexible skeleton to allow it to move fast. Cats can leap into the air, slink rapidly along the ground, and pounce like lightning on a tiny spot.

Strong muscles are also needed to move extra slowly. When it stalks, a cat has to move in "slow motion" so the prey will not notice it creeping up. To outwait a mouse, a cat often freezes as still as a stone for very long periods. Its strong muscles allow it to stay still for much longer than a person could.

The actions of a cat help to make it a good hunter too. Many of these actions are instinctive; that is, they have been inherited from the cat's wild ancestors and are built into its brain and nervous system. In a way, cats are "programmed" for mousing. They have automatic reactions to certain sights and noises that look and sound like small rodents.

For instance, all cats react to small moving objects by chasing them. Even kittens that have never hunted will chase anything that moves.

Cats also react instinctively to hidden scratching or rustling noises, which sound like mice moving in dry grass. And they will look into deep, round holes, which seem like the burrows of small mammals.

Obviously these instincts would help wild cats survive by putting them on the trail of small rodents nearby. But, in addition, such reactions make cats curious and playful, two traits that people seem to like in their pets.

Once the cat has located a mouse and is ready to capture it, still other instinctive reactions come into play. The cat goes through a series of actions that are always the same for every cat.

First the hunter creeps up on the prey, keeping its belly close to the ground. This movement is known as stalking. Then it crouches, waiting in ambush. It looks hard at the mouse, making sure the prey is in just the right spot. When it has the prey lined up, the cat pounces forward and grabs it with its forepaws. Unless it is hunting in long grass, the cat does not leap into the air. It keeps its hind legs firmly on the ground for balance.

The cat's natural way of hunting is to wait in ambush
and then pounce, keeping close to the ground.

Some bird lovers blame cats for killing large numbers of birds. But, while a few cats do become bird hunters, most can't catch birds at all. Cats evolved with the ability to catch small mammals that stay on the ground. They are not used to prey that can fly away. To kill birds, cats have to change their built-in hunting pattern.

They have to learn to stop creeping up on the prey and waiting in ambush, because while they are doing so, the bird often flies away. Instead, the cat must attack the bird as soon as it sees it. And to grab a bird as it begins to fly, a cat has to leap high instead of following its natural instinct to stay low on the ground.

For these reasons, most cats cannot hunt birds easily. Scientists say that out of every hundred animals killed by cats only about fourteen are birds. The rest are small mammals, mostly mice.

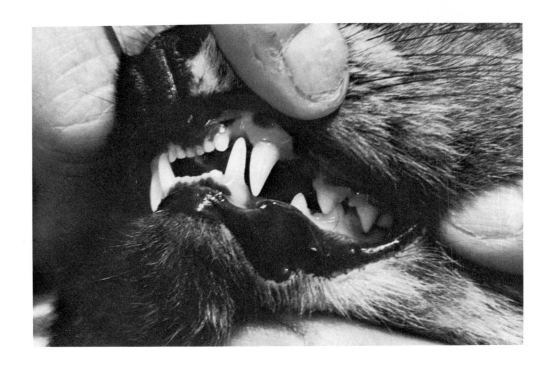

After catching the prey, the cat uses its four daggerlike canine teeth to make the kill. Again instinct plays a part in this phase of the hunt. The cat always delivers the killing bite to the same spot: the nape of the neck. The cat has a built-in reaction to the sight of the narrow place between the head and the body.

As the canine teeth bite, touch-sensitive nerves at their base guide them into the gaps between the prey's vertebrae (spine bones). These nerves are so accurate that the prey's bones are hardly ever bitten. Instead, the teeth force apart the vertebrae and kill the animal by biting through its spinal cord.

The cat has other kinds of teeth too. The twelve tiny teeth in front are the incisors. They are too small for biting and are used for scraping the last shreds of meat from bones.

Toward the rear of the mouth are fourteen large teeth known as molars. The molars of human beings and many other mammals are flat on top and are used for chewing and grinding. But cat molars are sharp and come together like scissor blades. Cats do not chew; instead, they use the molars for cutting meat into pieces small enough to swallow. For this reason, a cat bites a chunk of meat on the side of its mouth instead of in the front, as a human being would.

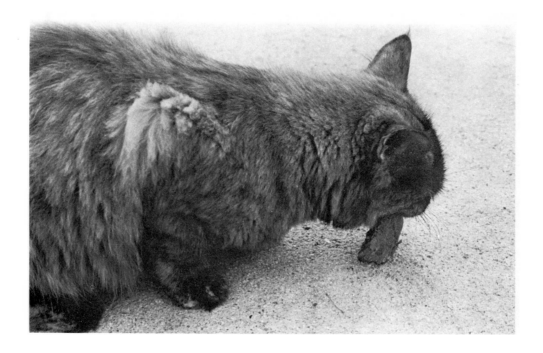

The cat's paws are also excellent tools for hunting.
The soft footpads and the fur between the toes give the
cat its silent walk, so necessary for hunting by stealth.

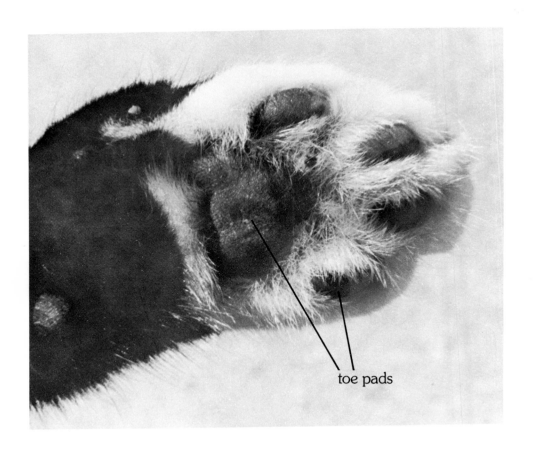

toe pads

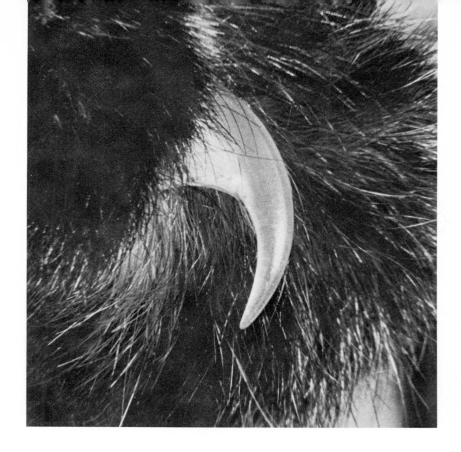

Hidden within the front paws are the cat's hooked claws, used for grabbing and holding prey. These claws are sharper than any other mammal's, and the cat keeps them that way by scratching them often on trees, furniture, or a post.

Usually the front claws stay safely inside the paws, where they are protected and stay sharp. When the cat wants to use them, it extends them with a system of muscles, tendons, and elastic ligaments. Claws that can go in and out this way are called "retractile."

The claws on the cat's back feet are not retractile, and the tips stick out of the fur. They are not as sharp as the front claws because they get worn down from walking, just as a dog's toenails do. A cat hardly ever uses its back claws as weapons, but if attacked by a larger animal, it will defend itself by rolling over on its back and clawing with all four feet.

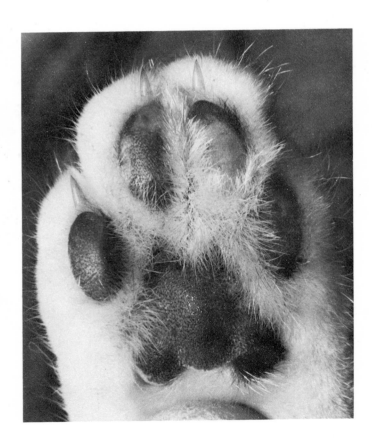

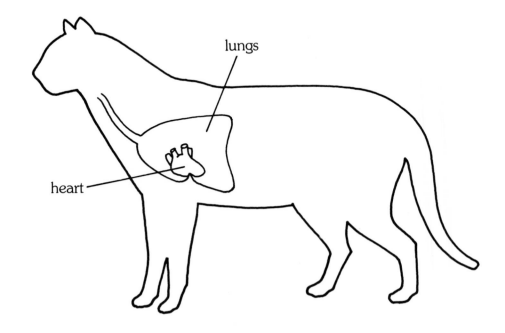

lungs

heart

Cats use great muscle power and energy when
hunting and defending themselves. However, because
their heart and lungs are small for their body size, they
tire easily. Therefore, cats are not usually lively for very
long periods. Instead, they spend time resting after
each spurt of activity.

Not only do cats rest a lot, they sleep twice as much as other animals. In many short catnaps throughout the day, cats sleep away two-thirds of their life. The reason why cats sleep so much remains a mystery to scientists.

A hunting animal needs keen senses for detecting prey, and the cat is no exception. The most important sense for a cat is its vision. Because cats' eyes are set forward on the face, they can judge well the distance and size of an object in front of them. They can also detect movement on the sides, but they have a large blind spot in the rear. As hunters, cats must be able to see prey and "take aim," but they have less need to detect attackers from behind.

By comparison, the eyes of prey animals, such as mice, are set on the sides of the head, and they can detect movement in almost a full circle.

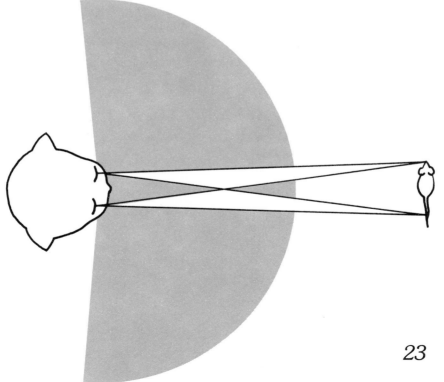

Cats do a lot of their hunting at night, and many people think that they can *see* in the dark. They cannot, of course; no animal can see in the total absence of light. However, cats *can see* in light six times dimmer than people can.

One reason for this excellent night vision is that the pupils of the *eyes* are very large. They can close to a tiny slit in bright sun, but open extra wide at night to let in as much light as possible.

26

Another reason cats see well at night is because of a special layer of cells at the rear of the eye. This layer acts like a mirror, reflecting back into the eye any light that was not absorbed the first time. Thus, the mirror "recycles" light and gives the cat's eye the chance to use every bit of it available.

This layer of mirror cells also makes cats' eyes glow in the dark if a bright light is flashed into the open pupils. The glow is sometimes called "night shine."

For hunting, cats rely first and foremost on their vision. But their second most important sense is their hearing. Cats can hear many sounds that are too faint or too high for the human ear. They are most sensitive to high-pitched sounds like the squeaks of kittens or mice.

The funnel-shaped outer ears can be turned to pinpoint sounds. In the pictures here, you can see how one ear swivels to pick up a sound to the right and rear of the cat.

The cat's sense of touch also tells it about its surroundings. The footpads and the hairless skin on the nose are sensitive to touch. You will often see a cat testing a new object by patting it with its paw and then by touching it with its nose.

The cat's whiskers are organs of touch too. These extra thick hairs grow on the upper lips, cheeks, chin, over the eyes, and on the inside of the forelegs.

The chin whiskers can feel a piece of food on the ground that the cat doesn't see. The lip whiskers can help a cat decide whether an opening is wide enough to squeeze through. And the foreleg whiskers can signal that a mouse is trying to escape the paws.

31

The cat has a good sense of smell, for there are about sixty-seven million cells inside its nose that pick up odors. Yet the cat does not use this sense very much in hunting. Instead, a cat's nose gives it information about other cats. Just by smelling, cats can tell if another cat is a stranger or one they know.

Cats and many other animals have an unusual sense organ in the roof of the mouth. It is known as Jacobson's organ, and it gives the animal a special sense that is a combination of taste and smell. Male cats use this sense to tell whether females are ready to mate. When a cat makes a strange face that looks like a grimace of disgust, it is actually testing an odor by pressing the tongue against the opening in the mouth that leads to Jacobson's organ.

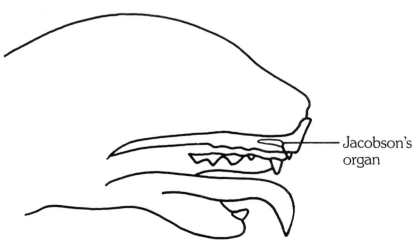

Jacobson's organ

A cat's body is protected from cold and wet by its sleek fur. The coat has two layers: the outer fur, which consists of longer, coarser hairs; and the underfur, which is soft and downy. The undercoat is what insulates it against cold, and when summer comes, a cat sheds it.

The cat has a special way of grooming its beautiful coat. With its rough tongue, it licks the fur clean. Occasionally it may bite out dirt with its tiny front teeth.

Because it is so limber, the cat can stretch to lick almost every part of its body.

The cat, however, cannot lick its face, head, and ears. To solve this problem, the cat first wets a paw and then rubs it vigorously over the hard-to-reach places.

One scientist believes that cats have a natural detergent in their saliva, because their fur smells so clean and sweet after grooming.

The cat's grooming ceremony, however, does more than keep its fur clean. The licking also stimulates the skin glands to produce more body oils for waterproofing the coat. And these oils contain a substance that is changed to vitamin D by sunlight. So when a cat washes, it swallows this valuable nutrient.

In addition, grooming helps cats stay cool in the summer. They produce watery sweat only on the footpads, so wetting the fur by licking is their substitute for sweating. When the saliva evaporates, it cools the cat's body.

Many people think of cats as loners, and in fact, except for lions, they do not live in groups. Each cat has its own territory, and each cat hunts by itself.

However, cats are not completely solitary animals. Even though they may stay alone in the daytime, they often gather at night for company. And cats that live in the same house as pets often become companions. They may play together, groom each other, and share the same resting places.

Whenever cats are together, they have ways of communicating with each other. They make faces to express feelings like anger, fear, and contentment. They also use body language. Switching the tail can mean "I am annoyed." Holding the tail straight up means "I am happy and friendly." Cats also "talk" to each other with sounds. They meow, hiss, growl, chirp. These noises can mean many things, from "hello" to "don't come any closer."

Cats' purring is another way of communicating. It tells other cats "I am content" and "all is well." No one knows exactly how cats purr, but many scientists think that the sound is caused by two folds of skin that lie in the cat's throat behind the vocal cords. When the cat breathes, these folds vibrate. But if so, scientists are still puzzled by the steady tone of the purr. Why doesn't it vary as the cat breathes in and out? No one knows.

A cat's facial expressions:

When a cat is *happy*, it perks its ears forward and relaxes its whiskers.

When *angry*, the cat lays back its ears and pushes its whiskers forward. The pupils of its *eyes* become small.

A *fearful* cat lays its ears and whiskers flat. Its pupils become large.

Another way that cats communicate with each other is through skin glands that give out scent. Cats have large numbers of these scent glands on the cheeks, the temples, and at the base of the tail. Although people cannot smell the scent, other cats can.

Cats use their scent to mark their territory. When a cat rubs a fence in its own yard with its cheeks and tail, it leaves a trace of its scent. This scent mark tells other cats "I was here" and "this is mine." It also reminds the cat itself "I belong here" and helps the cat feel at home.

Cats will even mark living things. When two cat friends meet, they sniff and rub against each other. The cat in the picture is actually marking its owner by rubbing the base of its tail against her legs.

Cats have other ways of marking too. Male cats
often spray urine on objects to mark them. And when
cats scratch trees or furniture, they may do so as
much for marking as for sharpening their claws. The
scratching leaves visible claw marks for other cats to see.

Because cats are naturally social, they make good pets. They are able to transfer to their human owners all the social skills that were originally meant for other cats.

A pet cat will communicate with a person using all the same meows and growls, facial expressions, and body talk that it would use with another cat. And human owners usually learn to understand their cat's language. They especially appreciate the cat's habit of purring. They seem to respond just as other cats do to the rumbling sound that signals all is well.

To survive as a species, cats, like other mammals, mate and bear young. The female cat comes into heat for several days a few times a year. Only during these times can mating take place.

After mating, egg cells (or ova) in the female may join with sperm cells from the male. Then a litter of kittens will begin to grow in the mother cat's womb.

In nine weeks, the kittens are born. They are helpless at birth and need their mother to keep them warm and clean and to give them milk.

As the kittens grow, they will learn to behave like cats by following their mother's example and by playing hunting and fighting games with each other.

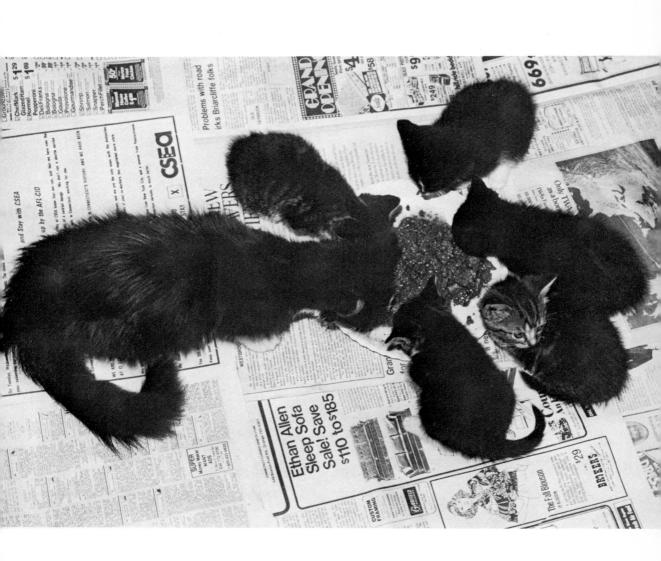

In only six weeks, the kitten is ready to leave its mother. In less than a year, it will be fully grown. Then the awkward bundle of fluff will have completed the change to a sleek adult. And whether or not it ever catches a mouse, it will always have the graceful body of a natural-born hunter.